Go Green

Green Living
Green Facts
Green Energy
And Tips For Going Green

By Ace McCloud
Copyright © 2013

Disclaimer

The information provided in this book is designed to provide helpful information on the subjects discussed. This book is not meant to be used, nor should it be used, to diagnose or treat any medical condition. For diagnosis or treatment of any medical problem, consult your own physician. The publisher and author are not responsible for any specific health or allergy needs that may require medical supervision and are not liable for any damages or negative consequences from any treatment, action, application or preparation, to any person reading or following the information in this book. Any references included are provided for informational purposes only. Readers should be aware that any websites or links listed in this book may change.

Table of Contents

Introduction 6
Chapter 1: Why Go Green? 7
Chapter 2: Go Green With How You Eat and Shop 9
Chapter 3: Go Green With How You Get Around 11
Chapter 4: Go Green At Home 13
Chapter 5: Go Green At Work 16
Chapter 6: Go Green With Energy 18
My Other Books and Audio Books 20

Be sure to check out my website for all my Books and Audio books.

www.AcesEbooks.com

Introduction

I want to thank you and congratulate you for buying the book, "Go Green: Green Ideas, Green Facts, Green Energy, Green Tips, and Green Projects To Make Your Life And The World A Better Place."

This book contains proven steps and strategies on how to "go green" in all aspects of your life, from what you eat and wear, to how you get from place to place, to recycling, green energy, and much more!

While some of the green ideas are large in scale and may require a significant initial monetary investment (that will eventually be recouped over time), most of the green suggestions in this book are things that do not require you to spend a large amount of money, and some cost you no money at all to implement. Nor does going green require an excessive amount of time. All that is required is that you have an interest in making your life and the world a better place, which in the long run can save you money, make you feel connected to your community, and give you the satisfaction of knowing that you did your part to help the planet.

Chapter 1: Why Go Green?

It should be common knowledge to everyone that over the last century or so, since the dawn of the Industrial Revolution, human activity involving the burning of fossil fuels has had an enormous environmental impact on the planet. Average temperatures are rising across the board, while extreme temperatures in both hot and cold areas are getting even more extreme. Snowstorms, monsoons, and hurricanes are becoming more severe, claiming more and more human lives. Icebergs in the poles are melting, causing sea levels to rise, and threatening large amounts of coastal developments and potentially untold sums of money and damage.

Climate change, then, is a problem that will affect all of humanity, regardless of race, class, religious or political preference, income level, etc. The changes to our way of life may not fully take hold for another generation or two, but sooner or later, the human race will have to own up to the fact that it is our behavior that is contributing to these catastrophic changes.

There are several reasons why you should want to go green. While you personally may not have to deal with the full extent of the environmental damage, your children, and certainly their children, will inherit a planet much more unstable than the one you inherited. Such environmental instability may increase food shortages, the likelihood of war, massive storms, increased pollution in the air, etc. All things that can contribute to a far more dangerous planet than the one we live in today.

At a much more local level, going green can bring a community together, giving people a common cause to join around, increasing political participation, and giving them a chance to know their fellow neighbors through outdoor gardening projects or just being outdoors more. While the race to save the planet from ourselves will have to be a global effort, it comes down to each and every community and person to join together and tackle climate change head on, in the process giving people a chance to feel more connected to the place they call home.

At the family level, going green can be a perfect opportunity for parents to bond with their children. Keeping up a summer garden, planting trees, building small green projects like a rainwater catcher, or even just walking to the store and talking instead of driving in the car and listening to the radio are activities that can bring people together for some positive human interaction. In the process, going green can teach children some valuable life lessons about personal responsibility, discipline, how local decisions affect people all across the globe, and much more. These lessons are important for kids to learn at a young age because it will prepare them to be the future leaders of their community.

Financially, it makes sense to go green. Not only does government at both the federal and state levels offer tax incentives for the average consumer for doing things such as purchasing a hybrid vehicle or having solar panels installed on the roof of your home, but those purchases will save you money in the long run through lower overall energy use. Large-scale purchases aside, even doing small things like maintaining a small vegetable garden at home to make salads, or resealing your windows and doors every year or two will save you money. The financial gains you will realize, coupled with the satisfaction of knowing you are doing your part to help the environment, will make the going green process all the more satisfying.

Chapter 2: Go Green With How You Eat and Shop

One of the easiest ways to go green is to vote with your wallet. Each time you go to the store to make a purchase, your money is sending a very clear message to the merchants, manufacturers, and service providers of the world economy about the items they are making available to you.

The simplest thing to do is pay attention to the label. Locally produced or locally sourced items have a much smaller carbon footprint, since they do not have to be shipped great distances in fossil fuel-burning trucks, trains, planes, or ships. Look for products that display an environmentally friendly label on the packaging. This lets you know that the item was produced and distributed in accordance with eco-friendly laws and regulations established by international governments and organizations.

Look for items that do not come in excessive wrapping, as packaging materials are one of the biggest contributors to landfills. Many stores will now put out items like nuts in big open containers, allowing each customer to scoop out what they want, eliminating a lot of individual packaging. If you can, try to bring your own containers with you and simply reuse them each time you go.

For the extremely motivated green consumer, consider calling or writing a letter or email to a company whose products you buy (or would like to buy), telling them how important it is to you and others that businesses try and do their part in helping save the planet. Customer input, no doubt, played a role in Nike's decision to organically manufacture all of its' products by 2011. The customer, after all, is always right, and if business owners see that being green is important to their customers, it will be important to them as well.

Big agricultural companies can be especially harmful to the environment, as they use massive amounts of fertilizers, which then run off into the water system when it rains. Smaller, family-run farms tend to farm using more organic methods, with little or no fertilizer use. Since their operations are smaller, the family farms tend to distribute their items closer to home, and therefore have much lower shipping distances, further lowering their carbon footprint. Check your local area for seasonal farmers' markets that will allow you to buy these locally grown and produced items, instead of always going to large-scale supermarkets that have long distribution chains.

When shopping, bring reusable bags with you so that you don't need to utilize the plastic bags that stores make available to you. If you are only purchasing one or two small items, politely comment to the cashier that you don't need a bag for your items and carry them out of the store instead. Leave your reusable bags in a place where you will see them each time you leave the house, such as by the front door.

Instead of buying plastic water bottles (which is usually just filtered tap water bottled in a far away state and shipped to you), think about purchasing a BPA-free water bottle and refilling it constantly. I personally like to use reusable metal water bottles. If you are concerned about the tap water cleanliness in your area, you can purchase a filter (either to attach to the faucet or a as a stand alone container) so that you can filter the water first and then refill your bottle. This would be a very green thing to do, since shipping plastic water bottles burns large amounts of fossil fuels and they create a tremendous amount of waste.

When going out to eat, try to support establishments that buy their food items locally. Most places that do source locally grown and produced items make a point of letting you know about it, either through an advertisement, a decal on the window, or on their menu. As more and more people buy into the idea of going green, more and more merchants will try to show that their business understands how important helping the environment is.

During the spring and summer months, it's a great idea to keep a vegetable garden in your yard, or even in an empty lot. Even if you just grow some salad greens and a couple of tomatoes, a garden is an easy and fun way to go green. Instead of going to the store to buy your produce items (which most likely came from far away), you can walk outside to your garden and gather what you need for your meal. You can go even greener by substituting chemical-based fertilizers and pesticides with organic compost and natural pesticides, which will be further discussed in Chapter 4. By having a garden, not only will your items be fresher and healthier, but it is also very fulfilling to grow your own food. And even if you don't have the yard space for a garden (if you live in an apartment, for example), even a couple of potted plants can do the trick. Salad greens and other herbs grow extremely well in window planters, and a small balcony is enough room to hold a couple of tomato pots so that you can enjoy fresh, local produce throughout the growing season.

If you do have the land available, along with a garden, you could consider planting some fruit trees and berry bushes. Many of the fruits you buy at the store, such as apples, for example, are shipped in from great distances, when they could be very easily grown right where you are. Fruit trees, like gardens, will require some amount of care, as you will have to prune them and spray them with herbicides and fungicides. When you do purchase any type of spray for your outdoor home area, always try to buy a natural or eco-friendly product. This way, your tree or veggie plants will remain chemical free, and your neighbors' yards will not get the chemical runoff that occurs after a heavy rainfall. And, in addition to providing you with fresh things to eat, fruit trees and berry bushes can serve other purposes. They can provide your home or patio area with shade during the hot summer months, or they can double as a natural fence to separate your property from a neighboring one. This way, you won't have to pay to have a fence installed or maintained.

Chapter 3: Go Green With How You Get Around

The easiest thing you can do to go green with regards to your mode of travel is start walking more. While having your own car and the ability to drive from place to place at your leisure might be very convenient, it is a very un-green way to move around. It's estimated that a single motor vehicle emits 8,000 pounds of carbon dioxide annually. Consider walking when possible (to the store, school, work, etc.), or carpooling. Many cities are going as far as building designated car pool lanes on highways, so that vehicles with more than one occupant can ride in a traffic-free lane.

Biking is also a green way to get around, and makes for great exercise. Many cities are realizing that their overly congested streets and highways are causing too much pollution, creating nightmare commutes, and unhealthy air conditions for its residents. Bike lanes and car-free downtown areas are slowly becoming the norm in metropolitan areas everywhere.

Making use of public transportation is also an easy way to go green. In rural areas, there often aren't enough people to support a mass transit system, so operating a private vehicle is usually the only realistic option to choose. But most large cities (and their surrounding suburbs) usually offer some combination of train/trolley and bus service, with buses sometimes running on eco-friendly natural gas. Even though it might be more convenient to drive yourself somewhere, keep in mind that commuters vote with their wallet also. The more people use the mass transit system, the more money municipalities will allocate towards expanding it, as opposed to pumping the money into car-friendly projects like highways and parking lots.

There are certain things you can do behind the wheel that will make your trip a bit more green as well. Fuel efficiency can be increased by as much as fifteen percent if you reduce your speed from sixty-five miles per hour to fifty-five miles per hour. Cutting it down from seventy to fifty-five will give you an even larger gain in fuel efficiency of twenty-three percent. While driving on a lightly-congested highway, use the cruise control feature to remain at a constant speed. Not accelerating and decelerating constantly will make your trip consume much less fuel than it would otherwise. Constantly check the air pressure of your tires to ensure that they are fully inflated. Deflated tires slow the vehicle down, demanding a greater amount of fuel to be used. Overall, be diligent about your oil changes and inspections so that a mechanic can catch any problem with your car right away and make the necessary repairs. Any car that is not running as efficiently as possible will always consume more gasoline than a finely-tuned one.

If you are in the market for a new car, consider buying a smaller vehicle, or even a hybrid. Large SUV's and trucks do fairly poorly when it comes to miles per gallon (in both the city and highway category) compared to smaller sedans, so going with the sedan would be helping out the environment a lot. But if you really want to go green with your vehicle, a hybrid is the way to go. Hybrids use much less fuel, as they often switch back and forth between electric and gas power. Some hybrids need to be charged while not in use, but charging stations are becoming more and more common, as manufacturers are realizing that simply relying on gas-powered vehicles to get around cannot last forever. And while hybrids are still more expensive than fully gas-powered vehicles of a comparable size with comparable features, governments at both the federal and state levels offer enticing tax breaks for making the switch to an eco-friendly car. Plus, in the long run, the money you save on gas with a hybrid will pay for the difference in price, and then some!

Chapter 4: Go Green At Home

Around the house, there are plenty of things you can do to make your home more green and lower its' overall carbon footprint.

Starting with the construction of the home, look into using green products, such as thermal insulation made from recycled material, instead of traditional construction products that will just add to landfill waste. Ask your contractor about incorporating bamboo products into the building process. More so than any other tree, bamboo consumes carbon dioxide at a much faster rate. For your kitchen, bathroom, and bedroom, consider buying used cabinets, couches, beds, dressers, etc. Additionally, tax credits are often available to consumers who purchase energy saving or environmentally friendly products.

If you have a lawn to maintain, using a push lawnmower is a great way to both go green and exercise at the same time. Traditional two-stroke lawnmowers pollute the equivalent of forty new cars with each hour of use, so utilizing a push lawnmower to cut your grass will definitely lower your home's carbon footprint. And if you have lots of landscaping to do (trees, flowers, etc.), think about using environmentally friendly products to treat your yard.

To reduce your household waste, be sure to check with your local sanitation department to see a list of all the possible items that can be recycled curbside or brought to the local dump. Most municipalities will now pick up paper and plastic materials, but for metal and glass items, you may have to bring them yourself during business hours.

To further cut down on household waste, you could start a compost pile if you have the available yard space. Composting food scraps and paper waste is a great way to lower the amount of garbage you put out each week, and also gives you fresh, organic soil to spread around your yard or use in a seasonal vegetable garden. Composting can be done either by digging a small ditch in the ground, or in some sort of container. However, if you use a container, be sure to puncture some holes in the sides so that water does not build up when it rains. Also, remember to alternate layers of wet material, such as food, with dry material, such as paper and cardboard. And once a week, use a pitchfork or shovel to turn the material several times in order to aerate it. The more air that is able to pass through the compost pile, the faster the material will decompose and turn into usable soil. Finally, if you live near a wooded area, you may not want to include meat scraps in your compost pile, as it may attract hungry animals such as raccoons and bears. Just stick with fruit, veggie scraps, and paper products.

At least once a year, take the time to check all of your exterior windows and doors, to make sure they are completely sealed and no air is escaping from inside of your home. If the caulk around the edge of the window or door has deteriorated, scrape it completely off and apply a new seal. This way, no cool air from your AC will escape in summer time, and now warm air from your furnace will escape in winter time, thereby maintaining a more stable interior temperature and lowering your overall cooling and heating costs.

During the summer months, try and remember to keep curtains and shades drawn during hours of intense sunlight. By blocking the direct sunlight, the rooms of your home will remain cooler, and your AC will not have to run as much. Planting trees or bushes around your property can also help block some sunlight during summer.

During the winter months, dress a bit more warmly while inside so that you don't have to keep the thermostat so high. While it may be more convenient and comfortable to wear shorts and a t-shirt, putting on long pants and a sweatshirt will allow you to lower your thermostat, save some money on your heating bill, and at the same time reduce your home energy requirements, which is great for the planet.

For handy individuals interested in going green, building a rainwater catcher is a cheap weekend project that can save you lots of money in the long run. The best place to position one is somewhere elevated, so that a tubing system can be installed to water your garden or flowers with the stored rainwater. But even an empty garbage can on the side of the house can do the trick. Just be sure to stir the water daily, as mosquitoes love to lay eggs in stagnant water.

When you encounter a dripping faucet in your home, repair or replace it immediately. National Geographic Kids estimates that around 2,700 gallons of water could be saved each year from households which have a faucet drip of one drop per second. A lot of times, fixing the drip is as simple as replacing a worn washer with a new one, which can cost less than a dollar to buy.

Look for detergents that are designed to be used with cold water instead of warm or hot water for washing clothes. Since such a large amount of energy is utilized to wash clothes in hot water, the average family that makes the switch can lower their carbon dioxide emissions by 1,281 pounds and save money as well.

Another way to cut back on your water use (and your water bill) is to install new low-flow showerheads, which are much more water-efficient than traditional showerheads. By doing so, you can experience as much as a thirty percent drop in water consumption and bills.

If your home is still equipped with incandescent light bulbs, head to the hardware store and replace them with compact fluorescent light bulbs (CFL's), which on average last ten times longer than traditional light bulbs. An added bonus of fluorescent bulbs is that they use up to seventy-five percent less energy.

If you have the money and are looking to make some long-term investments in your home that will significantly raise the property value, there are many green projects you can have done. These include installing solar panels on the roof (which can even turn into a money-making venture, as excess electricity can sometimes be sold to the local utility company), putting in brand new energy-efficient appliances (stove, washing machine, refrigerator, etc.), installing tight-sealing interior doors and individual thermostats in each room (which may require updating your current HVAC system), and replacing all exterior windows with energy-efficient double-paned windows that do a much better job of insulating your home. As with building materials, there are a host of tax breaks for green-minded consumers to take advantage of.

Chapter 5: Go Green At Work

There are plenty of easy and effective ways to go green at work. As mentioned previously, carpooling or taking public transportation will cut down on the carbon emissions required to get you to and from your job, and save money on gas as well.

During the summer months, ask about being allowed to dress more casual instead of having to wear a suit and tie (or other long-sleeved outfit). This way, the AC won't need to be run as much, which not only saves the company money, but reduces energy requirements as well.

If telecommuting is an option, take advantage of working from home from time to time to cut down on your commute altogether. With just a computer, multi-purpose printer, phone, and internet connection, anyone can set up a home office.

Instead of using a different cup for coffee and water each time, keep a coffee mug and water bottle at work to cut down on waste. Do the same with plates and silverware if you tend to bring your meals from home instead of going out. While at times it may be easier just to buy a new bottle every day, keep in mind that more than twenty-five million plastic bottles are discarded every hour. Plastic and styrofoam material usually end up in the ocean, and contribute to contaminating the water and causing problems for sea creatures.

When an ink or toner cartridge is empty, recycle it instead of throwing it in the trash. Cartridges can easily be refilled and prepared for reuse, thereby cutting back on landfill waste. According to the office supply chain Office Depot, two and a half pounds of metal and plastic and half of a gallon of oil goes into making each ink and toner cartridge. Simply by recycling these items you can do an immense amount of good for the environment.

Paper is another big waste item in the office. According to Resourceful Schools, the equivalent of 185 gallons of gas is saved for every one ton of mixed recycling paper. Also worth mentioning is that, according to the same organization, thirty-six percent of the fiber utilized in the production of new paper products is derived from recycled paper. To cut back on paper waste, consider reading a newspaper online in electronic format. If everyone did so, it is estimated that 500,000 trees would be saved each week.

Junk mail is also a huge contributor to both office and home waste. Huge amounts of junk mail are created each year. Try to cut down or recycle as much junk mail as possible.

When not in use, turn off power strips and unplug appliances that do not need a constant flow of energy to operate. Adjust the settings on computers so that after a period of inactivity, the computer automatically goes into an energy-saving mode without completely shutting off.

For cleaning the office, it's a good idea to use biodegradable cleaning supplies or detergents made of natural products. Chemical-heavy products eventually end up in the sewer system, and a lot of energy must be expended to treat the water before it is ready for human use again.

When a large electronic device such as a computer, printer, or television is ready to be replaced with a newer model, recycle it instead of placing it in the trash. Many non-profit organizations will gladly take such items off your hands and distribute them to schools and community centers desperate for any type of equipment. Another reason to reuse or recycle electronic equipment is that, if placed in a landfill, mercury and several other harmful toxins can seep out into the ground, contaminating both it and the local water supply (this according to the World Watch Institute).

Recycling metal is one of the easiest things you can do at work to go green. Every forty-five seconds a quarter of a million aluminum cans are manufactured, producing large amounts of metallic waste as a by-product. Simply placing a recycling container next to the soda machine will provide workers with an easy and environment friendly way to dispose of their can. These cans are then reused to make other metallic products, saving energy. Enough energy, in fact, that for every soda can recycled, enough energy is saved to watch three hours worth of television.

Glass is another material that can easily be recycled to avoid ending up in landfills. Considering that a single glass bottle takes about four thousand years to decompose, you should try and recycle glass as much as possible. Fortunately, unlike plastics, glass can be recycled an unlimited number of times. Given that a good amount of the forty-one billion glass containers that are manufactured each year are thrown away, and that a one hundred-watt light bulb can be powered for three to four hours just by recycling one glass bottle, it is in the best interest of anyone who cares about the planet to do their part and recycle as many glass products as they can.

Chapter 6: Go Green With Energy

Perhaps the most effective thing you can do to go green is to switch to a utility company that supplies some or all of your energy from green energy sources. Equally as effective is using items like solar panels and home wind turbines, to simply harness the energy yourself. Currently, most of the energy in the United States and abroad comes from non-renewable sources such as coal and oil. Not only is the burning of these fossil fuels harmful to the environment, but their scarcity and high market prices have been the cause of massive instability in areas such as the Middle East. This being the case, humans have finally accepted the fact that they need to seek out alternative fuel sources that will be better for the planet in the long run.

The most obvious alternative form of energy comes from the sun. Solar energy, once harnessed using solar panels, can be used for residential, commercial, and industrial uses. The downside of solar energy is that panels are still a bit pricey, however, they can be used quite effectively during the colder times of the year, as long as it is sunny, something a lot of people don't know. Solar energy may be more effective for homeowners in sunny climates, since they will tend to generate more power from their solar panels, and they can sell excess electricity back to the power company at a profit in participating locations. By this method, coupled with generous government tax breaks, the panels will pay for themselves and provide a steady source of income in the future.

Wind energy can be captured efficiently using large or small wind turbines, which are perfect for mounting to the top of your home or other tall structures (such as a standalone tower). When the wind blows, the device's elongated propellers (similar to an airplane's) spin, generating electricity. The electricity then enters the power grid and is moved to a needy market. The downside of the turbines is that could break down or require maintenance. And, it goes without saying that if the wind stops, so does the power production. But, If you are in a windy place, wind turbines should definitely be looked into!

Fortunately, both wind turbines and solar panels can be connected directly to the power grid, with captured generated electricity being pumped directly into your home. Also, solar panels and wind turbines can be attached to batteries that can provide power in cases of a power outage. These battery backup power systems can range from just a few batteries to power a few small items, to up to 24, 50, or more batteries that could power a house for a week or more in a green environment. A much better alternative to gas powered generators in many ways. Not only do batteries make no noise, but they don't emit fossil fuels either.

Conclusion

I hope this book was able to help you understand why going green is important, and give you plenty of ideas on how to go green in all aspects of your life.

The next step is to start going green. Start with small steps, and then work your way up to big green projects. There is nothing quite as satisfying as knowing that you are doing things right and that are beneficial to the world as a whole!

Finally, if you discovered at least one thing that has helped you or that you think would be beneficial to someone else, be sure to take a few seconds to easily post a quick positive review. As an author, your positive feedback is desperately needed. Your highly valuable five star reviews are like a river of golden joy flowing through a sunny forest of mighty trees and beautiful flowers! *To do your good deed in making the world a better place by helping others with your valuable insight, just leave a nice review.*

My Other Books and Audio Books
www.AcesEbooks.com

Peak Performance Books

Health Books

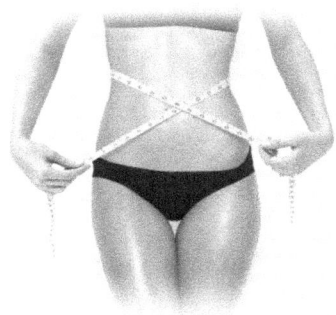

Be sure to check out my audio books as well!

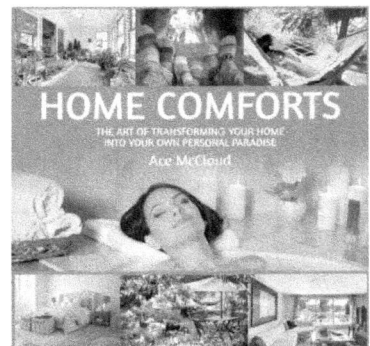

Check out my website at: **www.AcesEbooks.com** for a complete list of all of my books and high quality audio books. I enjoy bringing you the best knowledge in the world and wish you the best in using this information to make your journey through life better and more enjoyable! **Best of luck to you!**

www.ingramcontent.com/pod-product-compliance
Lightning Source LLC
Chambersburg PA
CBHW051432070526
44584CB00023B/3688